JAN 16 2004   Doj

DISCARDED
From Nashville Public Library

D1467935

Property of
Nashville Public Library
615 Church St., Nashville, Tn. 37219

LIFE CYCLES
Beans

by Melanie Mitchell

first step nonfiction

Lerner Publications Company · Minneapolis

These are **beans.**

There are many kinds of beans.

How do beans grow?

Beans are seeds.

A bean seed is planted.

First the seed grows **roots.**

Next a **shoot** begins to grow.

Now it is a **seedling.**

The seedling grows bigger.

The young bean plant grows
flowers.

Seeds grow inside the flowers.

Pods grow around the seeds.

The seeds are beans. Some
pods are also beans.

These beans are ready to
be picked.

People will eat some of the beans.

The rest will be planted to grow new bean plants.

Life Cycle of a Bean

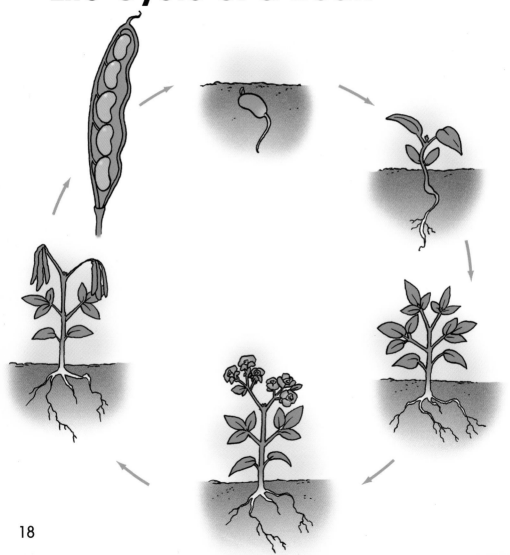

Beans

Beans have been around for thousands of years. Beans are grown all over the world. Different types of beans grow in different places. North Dakota and Michigan grow the most beans in the United States.

The life cycle of a bean is between 120 and 150 days. This means beans planted in the spring are ready to be picked in late summer or early fall.

Bean Facts

January 6 is National Bean Day.

Green beans have been grown in Mexico for at least 7,000 years.

It takes 40 to 45 beans to fill one can of cut green beans.

The Mexican Jumping Bean is not a bean. It is a seed that contains the larva of a moth called the Jumping Bean Moth.

Some crayons are made from soybeans.

Lima beans are named for Lima, Peru.

Boston, Massachusetts, is nicknamed Beantown. It is famous for a dish called Boston baked beans.

Chocolate is made from the cacao bean. Cacao beans grow in warm, tropical places.

Glossary

 bean – seed or pod used for food

 pods – parts of a plant that protect seeds

 roots – parts of a plant that grow down into the soil

 seedling – a young plant

 shoot – a plant that has just started to grow

Index

Copyright © 2003 by Lerner Publications Company

All rights reserved. International copyright secured. No part of this book may be reproduced, stored in a retrieval system, or transmitted in any form or by any means—electronic, mechanical, photocopying, recording, or otherwise—without the prior written permission of Lerner Publications Company, except for the inclusion of brief quotations in an acknowledged review.

The photographs in this book are reproduced through the courtesy of: © Dwight R. Kuhn, front cover, pp. 2, 5, 6, 7, 8, 9, 10, 12, 13, 14, 22 (all); © Walt and Louiseann Pietrowicz/September 8th Stock, p. 3; © CORBIS Royalty Free Images, p. 4; USDA Photo, pp. 11, 15, 17; © Robert Maust/Photo Agora, p. 16.

Illustration on page 18 by Tim Seeley.

Lerner Publications Company,
A division of Lerner Publishing Group
241 First Avenue North
Minneapolis, MN 55401 USA

Website address: www.lernerbooks.com

Library of Congress Cataloging-in-Publication Data

Mitchell, Melanie S.
    Beans / by Melanie Mitchell.
        p.   cm. — (First step nonfiction) (Life cycles)
    Summary: A basic overview of the life cycle of a bean plant.
    Includes information about different kinds of beans and their uses.
    ISBN: 0–8225–4608–6 (lib. bdg. : alk. paper)
    1. Beans—Life cycles—Juvenile literature. [1. Beans.] I. Title.
    II. Series.
    SB327 .M58 2003
    635'.652—dc21   2002004714

Manufactured in the United States of America
1 2 3 4 5 6 – JR – 08 07 06 05 04 03